| 中英对照 |

中国石油
二氧化碳捕集、利用与封存大事记

CHRONICLE OF MAJOR EVENTS ON CCUS IN CNPC

中国石油天然气集团有限公司碳中和技术研发中心 编

Edited by R&D Center for Carbon Neutralization Technology of CNPC

石油工业出版社

Petroleum Industry Press

内 容 提 要

本书以中英文对照的方式详细总结和梳理中国石油天然气集团有限公司在二氧化碳捕集、利用与封存技术实践的一系列战略举措、突破性进展、标志性成果等大事件纪实。

本书可供石油企业广大员工以及关心能源行业发展的各界读者参考。

图书在版编目（CIP）数据

中国石油二氧化碳捕集、利用与封存大事记/中国石油天然气集团有限公司碳中和技术研发中心编. —北京：石油工业出版社，2024.6

ISBN 978-7-5183-6327-8

Ⅰ. ①中… Ⅱ. ①中… Ⅲ. ①石油工业－二氧化碳－收集－研究 Ⅳ. ① X701.7

中国国家版本馆 CIP 数据核字（2023）第 190840 号

出版发行：石油工业出版社
（北京安定门外安华里 2 区 1 号　100011）
网　　址：www.petropub.com
编辑部：（010）64523561
图书营销中心：（010）64523633
经　　销：全国新华书店
印　　刷：北京中石油彩色印刷有限责任公司

2024 年 6 月第 1 版　2024 年 6 月第 1 次印刷
889×1194 毫米　开本：1/16　印张：4.5
字数：200 千字

定价：98.00 元
（如出现印装质量问题，我社图书营销中心负责调换）
版权所有，翻印必究

《中国石油二氧化碳捕集、利用与封存大事记》

编写委员会

顾　　问：孙龙德　袁士义　沈平平　江同文
　　　　　周爱国　钟太贤　赵邦六　杜卫东
主　　编：雍瑞生
副 主 编：陈宏坤　袁　波　赵兴雷　史　方　李巨峰
成　　员：张坤峰　蔡明玉　张德平　吴百春　高　明　杨永智　陈昊卫
　　　　　陈情来　刘知鑫　孙晋平　白文广　雷友忠　王靖华　范　伟
　　　　　董海海　林贤莉　李　清　李明卓　刘安琪　那慧玲　杨术刚
　　　　　杨川箬　叶　舣　刘双星　吴慧君　吴　剑

Chronicle of Major Events on CCUS in CNPC

Editorial Board

Consultants:	Sun Longde	Yuan Shiyi	Shen Pingping	Jiang Tongwen
	Zhou Aiguo	Zhong Taixian	Zhao Bangliu	Du Weidong
Editor in Chief:	Yong Ruisheng			
Deputy Editor in Chief:	Chen Hongkun	Yuan Bo	Zhao Xinglei	Shifang
	Li Jufeng			
Members:	Zhang Kunfeng	Cai Mingyu	Zhang Deping	Wu Baichun
	Gao Ming	Yang Yongzhi	Chen Haowei	Chen Qinglai
	Liu Zhixin	Sun Jinping	Bai Wenguang	Lei Youzhong
	Wang Jinghua	Fan Wei	Dong Haihai	Lin Xianli
	Li Qing	Li Mingzhuo	Liu Anqi	Na Huiling
	Yang Shugang	Yang Chuanruo	Ye Yi	Liu Shuangxing
	Wu Huijun	Wu Jian		

序

　　二氧化碳捕集、利用与封存（CCUS）作为二氧化碳减排的有效途径，是实现"双碳"目标的兜底技术。可以预见，CCUS 在全球气候治理中必将发挥越来越重要的作用。

　　回顾 CCUS 的发展历程，中国早在 20 世纪 60 年代就开始了探索和试验。1965 年，大庆油田 CO_2 注入与埋存试验无可争议地是我国 CCUS 的先驱。后来，中国石油又牵头依托国家重大科技专项、973 和 863 等项目，持续开展攻关、示范，一直引领着该领域的工作，成为行业的领跑者。据统计，截至 2022 年底，全国已投运和规划建设中的 CCUS 示范项目近百个，具备二氧化碳年捕集能力约 400 万吨，年注入能力约 200 万吨，其中中国石油已累计注入 563 万吨、年注入量首次突破 100 万吨，分别约占全国已注入量的 70% 以上和年注入量的 50% 左右。

　　由中国石油天然气集团有限公司碳中和技术研发中心编撰的《中国石油二氧化碳捕集、利用与封存大事记》（以下简称《大事记》），通过梳理 50 多年来中国石油 CCUS 发展历程，展现了中国石油稳步打造 CCUS 大规模产业链的努力和成果。从 1965 至 2006 年为探索起步阶段，首先是以大庆油田室内实验为基础，选择长垣区块开展井组规模碳酸水试注，迈出国内 CCUS-EOR 技术探索的第一步；经过四十载的探索研究，2006 年矿场试验规模逐步扩大，并在香山会议首次提出中国发展 CCUS 产业倡议。2007—2021 年为攻关试验阶段，以承担国家 973 项目为标志，大面积推进工业先导试验，并开展了更广泛的国际合作。2022 年及以后为工业示范阶段，CCUS 项目在驱油利用方向正式迈入工业化应用，中国石油已基本形成了二氧化碳捕集、输送、驱油与封存全流程配套技术，走出了一条由重驱油向驱油和减碳并重的全产业链一体化发展的 CCUS 产业发展之路。《大事记》向读者描绘了中国石油 CCUS 产业历经理论探索、技术攻关与试验、工业化试验与推广三个阶段不同寻常的画面，展示了中国石油先后在松辽盆地、鄂尔多斯盆地等区域布局多项 CCUS 重大工程示范和先导试验项目，以及在技术创新、注气规模、组织管理等方面取得的显著成效。另外，作者以展望的方式突出了中国石油将绿色低碳纳入公司发展战略，确定新能源业务"三步走"总体部署，并将 CCUS 示范工程作为绿色产业的"六大基地、五大工程"之一重点布局；规划了 2025 年、2035 年和 2050 年三个时间节点中国石油 CCS/CCUS 产业发展目标。

"双碳"目标，关乎当前、影响未来，是构建人类命运共同体的崇高事业。CCUS 与"双碳"目标息息相关，事关能源转型与能源革命，我们希望在这条通往未来的大道上，你追我赶，万马奔腾，为我们的地球创造出一个绿色宜居的美好明天。

中国工程院院士 孙龙德

Foreword

Carbon Capture, Utilization, and Storage (CCUS), as an effective approach for carbon dioxide (CO_2) emission reduction, represents an essential technology in achieving the carbon peaking and carbon neutrality goals. It is foreseeable that CCUS will increasingly play a crucial role in global climate governance.

Looking back at the Development of CCUS, China initiated exploration and experimentation in Carbon Capture, Utilization, and Storage (CCUS) as early as the 1960s. The indisputable pioneer of CCUS in China was the CO_2 injection and storage trial at the Daqing Oilfield in 1965. Subsequently, CNPC took the lead in leveraging major national scientific initiatives such as the National Science and Technology Major Project, 973 Program, and 863 Program to continuously conduct research, development, and demonstration activities, consistently leading the efforts in this field and emerging as an industry frontrunner. As of the end of 2022, statistics showed that nearly 100 CCUS demonstration projects were either operational or in the planning and construction phases nationwide. These projects collectively possess an annual CO_2 capture capacity of approximately 4 million tons and an annual injection capacity of around 2 million tons. Notably, CNPC has cumulatively injected 5.63 million tons of CO_2, marking the first-time annual injection volume surpassing 1 million tons. These figures respectively represent over 70% of the nation's total injected volume and approximately 50% of the annual injection rate.

Compiled by the Carbon Neutrality Technology Research and Development Center of CNPC, *Chronicle of Major Events on CCUS in CNPC* (hereinafter referred to as the Chronicle) reviews CCUS development of CNPC over the past 50-plus years, presenting its steadfast efforts and achievements in building a large-scale CCUS industrial chain. The period from 1965 to 2006 was an exploratory phase in its endeavor, commencing with indoor experiments at the Daqing Oilfield and advancing to pilot-scale well group carbonated water test injection in the Changyuan block, representing the initial step in the domestic exploration of CCUS-EOR technology. After four decades' research and exploration, the scale of field trials gradually expanded, culminating in the proposition of China's CCUS industry development initiative at the Xiangshan Conference in 2006.

The period from 2007 to 2021 constituted the research and experimental phase, epitomized by the undertaking of the National 973 Program, driving large-scale industrial pilot trials and fostering broader international collaboration. The post-2022 period signifies the industrial demonstration phase, where CCUS projects officially transition towards industrial applications in enhanced oil recovery. CNPC has essentially developed a complete set of technologies encompassing the capture, transportation, EOR and storage of CO_2, paving the way for an integrated CCUS industry development roadmap with the focus shifted from EOR to both EOR and carbon reduction. The Chronicle vividly depicts the extraordinary journey of CNPC's CCUS industry through three distinct phases: theoretical exploration, technological breakthroughs and experimentation, trial and promotion of its industrialization. It showcases CNPC's significant achievements in the layout of several major CCUS engineering demonstration and pilot projects in regions such as the Songliao Basin and Ordos Basin, as well as notable progress in technological innovation, injection scale, and organizational management. Additionally, the author highlights the following content in anticipation: CNPC has included green and low-carbon initiatives into its corporate development strategy, outlined a three-step overall deployment plan for new energy businesses, and positioned CCUS demonstration projects as a key focus within the framework of the "Six Major Bases and Five Major Projects" for green industries; furthermore, the author delineates CNPC's development goals in CCS/CCUS for the years 2025, 2035, and 2050.

The carbon peaking and carbon neutrality goals, bearing significance in the present and shaping the future, represent a noble endeavor towards building a community with a shared future for mankind. CCUS is intricately linked to the carbon peaking and carbon neutrality goals, pivotal in energy transition and revolution. As we embark on this path towards the future, let us strive together to create a green and habitable world for our planet.

Academician of Chinese Academy of Engineering

前 言

2020年9月，习近平主席在第七十五届联合国大会上向全世界宣布，中国二氧化碳排放力争于2030年前达到峰值、努力争取2060年前实现碳中和。这彰显了我国积极应对气候变化、构建人类命运共同体的决心与大国担当。党中央关于碳达峰、碳中和的重大战略决策，是我们对国际社会的庄严承诺，多项重大举措的部署为我国绿色、低碳、高质量发展指明方向。二氧化碳捕集、利用与封存技术（CCUS）作为实现碳排放变废为宝、效益与环保并重的一项绿色开发技术，在助力我国实现"双碳"目标的同时，可以大幅提高石油低品位资源和煤基能源的开发利用率，为保障国家能源安全提供支撑。近年来，在国家系列政策的支持推动下，CCUS技术呈现出良好的发展势头。随着全球应对气候变化进程的加速推进，CCUS作为目前唯一能够实现化石能源大规模低碳化利用的减碳固碳技术，已成为大多数国家碳中和行动计划的重要组成部分。

作为中国最大的油气生产供应企业，中国石油天然气集团有限公司（以下简称中国石油）牢记"我为祖国献石油"的初心使命，持续深化改革、强化创新，大力推进绿色低碳转型，致力于成为实现国家"双碳"目标和保障国家能源安全的中坚力量。中国石油党组高度重视科技创新和绿色低碳发展，将两者纳入公司发展战略，成立"双碳三新"领导小组，确定了绿色低碳"三步走"战略部署，制订碳达峰行动方案和科技支撑"双碳"行动方案，将CCUS作为其中的重要内容，成立CCUS工作专班，综合性布局了一系列重大示范工程和重大科研项目。2022年6月5日，中国石油发布《中国石油绿色低碳发展行动计划3.0》，将"CCUS产业链建设工程"作为迈向碳循环经济的十大工程之一，并设定2025年中国石油CCUS/CCS产业链建设将形成五百万吨级注入能力，到2035年将达到亿吨级规模，到2050年形成引领CCUS产业发展能力的目标。中国石油打造CCUS产业链的宏伟蓝图已徐徐展开。

回首中国石油 CCUS 领域五十多年成长过程，几代石油人不断深耕厚植和持续攻坚，稳步扩大应用规模，形成了成套技术体系和初具规模的产业示范。早在 20 世纪 60 年代，大庆油田就进行了二氧化碳利用的探索试验，拉开了我国二氧化碳捕集、驱油封存（CCUS-EOR）的序幕；20 世纪 90 年代初，吉林油田启动了 CCUS-EOR 全产业链技术攻关、应用试验，建成全国首个全流程二氧化碳捕集与驱油示范工程。21 世纪初，为进一步推动在 CCUS 理论技术、现场试验、工业化示范应用等全产业链发展，中国石油依托国家 973 计划项目、863 计划项目和重大科技专项等持续开展科技攻关、示范，突破多项关键核心技术，并强化国际交流与合作，参与国际标准制定，提升了我国在此领域的影响力。2020 年以后，中国石油推动 CCUS 的步伐大大加快，成立了 CCUS/CCS 重点实验室、碳中和技术研发中心；设立了 CCUS 重大科技专项，加强关键核心技术攻关；牵头组建了全国 CCUS 标准化工作组，并积极参与国际标准制定工作；启动 OGCI 昆仑气候投资基金，重点投资 CCUS 等温室气体减排技术。在加强核心技术攻关的同时，中国石油持续推动 CCUS 产业链建设及工程示范，在吉林油田和长庆油田建成了两个 CCUS 国家级示范工程，部署以松辽盆地 300 万吨 CCUS 示范工程为代表的"四大工程示范"和"六个先导试验"，制订两个千万吨级 CCUS 中长期发展规划。2022 年，中国石油已实现 111 万吨二氧化碳注入、30 万吨以上年产油能力，累计注入二氧化碳 563 万吨，占全国已注入量的 70% 以上。

本书首次系统梳理了中国石油五十多年来在 CCUS 领域的发展历程，总结了中国石油 CCUS 事业的一系列战略性举措、突破性进展、标志性成果，为关心能源行业 CCUS 技术发展的朋友提供了丰富、翔实、可靠的资料。回首，是为了更好的出发。我们深信，中国石油的 CCUS 之路将会越走越宽广，在中国实现碳中和的史册中留下浓墨重彩的篇章。我们编写中国石油 CCUS 大事记，也是为了回顾过往，总结经验，更好地推动中国石油绿色低碳转型，为我国深入推进能源革命、加强煤炭清洁高效利用、加快建设新型能源体系、积极参与应对气候变化全球治理等重大决策部署做出积极贡献。

本书的编写得到了中国石油各企业、研究院所的大力支持，为编写委员会提供了大量基础资料，在此表示衷心的感谢。如有疏漏之处竭诚欢迎批评指正。

Preface

In September 2020, Chinese President Xi Jinping announced at the General Debate of the 75th session of the United Nations General Assembly that China will strive to have carbon dioxide emissions peak by 2030 and achieve carbon neutrality by 2060. This demonstrates China's determination and responsibility as a major country to actively address climate change and build a community with a shared future for mankind. The decision on carbon peaking and carbon neutrality is China's commitment to the international community, and the deployment of a number of major measures has pointed out the direction for China's green, low-carbon and high-quality development. As a green development technology that translates carbon emissions into treasure while achieving benefits and environmental protection, carbon dioxide capture, utilization and storage (CCUS) technology can greatly improve the development and utilization rate of low-grade oil resources and coal-based energy while helping to achieve China's "dual carbon" goal, and provide support for ensuring national energy security. In recent years, with the support of a series of national policies, CCUS technology has shown a good momentum of development. As the global response to climate change moves forward, CCUS, the only carbon reduction and sequestration technology that can realize scalable low-carbon utilization of fossil energy, has become an important part of carbon neutrality action plans in most countries.

As China's largest oil and gas producer and supplier, CNPC always keeps in mind its mission to dedicate oil to the country. We have made sustained efforts to deepen reform, strengthen innovation, and advance green and low-carbon transition. We are committed to becoming the pillar in achieving the national "dual carbon" goal and ensuring national energy security. We highly value technological innovation and green & low-carbon development, and have incorporated them into our development strategies. We have established a leading group of "double carbon + new energy, new materials, new economy", rolled out a three-step roadmap for green and low-carbon development, and developed a carbon peak action plan and a "double carbon" technology support action plan. We have set up a CCUS working group and deployed a series of major engineering and scientific research project clusters. On June 5, 2022, CNPC released *CNPC Green and Low-Carbon Development Action Plan 3.0*, taking the "CCUS industry chain construction project" as one of the ten major projects towards a carbon circular economy, and setting a goal to build an injection capacity of five million tons in 2025 and 100 million tons in 2035 and lead the development of CCUS/CCS industry in 2050. CNPC has slowly unfolded its blueprint for building a large-scale CCUS industrial chain.

With more than fifty years of sustained efforts in the field of CCUS, we have expanded the scale of application steadily and established a complete set of technical systems, with industrial demonstrations having taken shape. As early as the 1960s, Daqing Oilfield carried out exploratory tests on the utilization of CO_2, which kicked off the prelude to carbon dioxide capture, flooding, and storage (CCUS-EOR) in China. In the early 1990s, Jilin Oilfield launched the technical research and application test of the whole industrial chain of CCUS-EOR, and established the first full-process carbon dioxide capture and flooding demonstration project in the country. At the beginning of the 21st century, to further promote the development of the whole industrial chain in CCUS theory and technology, field tests, industrial demonstration applications, etc., we have made continuous efforts in technology research relying on the National 973 Program, 863 Program, and other major science and technology projects. We have made breakthroughs in a number of key core technologies, strengthened international exchanges and cooperation, and participated in the formulation of international standards, which has enhanced China's influence in this field. Since 2020, to support the national "dual carbon" goals, CNPC has stepped up its efforts to promote CCUS and established the CCUS/CCS Key Laboratory, the Carbon Neutral Technology R&D Center. We have launched CCUS major science and technology projects to enhance the research of key core technologies. We have led the establishment of the National CCUS Standardization Working Group, and actively participated in the formulation of international standards. We have initiated the OGCI Kunlun Climate Investment Fund, focusing on investing in greenhouse gas reduction technologies such as CCUS. While strengthening the research of key core technologies, we have also made sustained efforts to advance the construction and demonstration of the CCUS industry chain. We have built two national CCUS demonstration projects in Jilin Oilfield and Changqing Oilfield, deployed four major project demonstrations and six pilot tests represented by the 3-million-ton CCUS major demonstration project in the Songliao Basin, and developed a medium-and long-term development plan for building two 10-million-ton CCUS industrial bases. In 2022, we achieved a carbon dioxide injection of 1.11 million tons, an annual oil production capacity of more than 300,000 tons, and a cumulative carbon dioxide injection of 5.63 million tons, accounting for more than 70% of the country's injected volume.

This book presents milestones of CNPC in the field of CCUS over the past fifty years and provides a summary of CNPC's strategic measures, breakthroughs and landmark achievements in CCUS business, which can be served as a reference for people who are concerned about the development of CCUS technology in the energy industry. Looking back is for a better start. We are convinced that CNPC will see a promising future for CCUS and leave a mark on China's road to carbon neutrality. We hope that with this chronicle, we can move further forward with CNPC's green and low-carbon transition, and make positive contributions to China's major decisions such as advancing the energy revolution, strengthening the clean and efficient use of coal, speeding up the planning and construction of a new energy system, and actively participating in the global governance of climate change.

During the writing process of this book, we have received great support from companies and research institutes under CNPC for the compilation of this chronology. We would like to express our sincere gratitude to them for the wealth of information provided. Correction is welcome if any error is found.

目 录
CONTENTS

1 中国石油 CCUS 历程
Development History of CCUS in CNPC 01

2 中国石油 CCUS 大事记
Chronicle of Major Events on CCUS in CNPC 06

   2.1 探索阶段（1965—2006 年）
      1965–2006: Exploration Stage 07

   2.2 攻关试验阶段（2007—2021 年）
      2007–2021: Technical Research and Test Stage 13

   2.3 工业示范阶段（2022 年及以后）
      2022 and Beyond: Industrial Demonstration Stage 35

3 中国石油 CCUS 展望
Outlook on CCUS in CNPC 44

附表 1　中国石油 CCUS 大事记一览表 47
Appendix 1　Table of Chronicle of Major Events on CCUS in CNPC ... 49

附表 2　中国石油开展的 CCUS 相关研究项目 51
Appendix 2　CCUS-related Research Projects Undertaken by CNPC ... 52

附表 3　中国石油 CCUS 主要示范工程 53
Appendix 3　Major CCUS Projects of CNPC 54

1 中国石油 CCUS 历程

1 Development History of CCUS in CNPC

1965 年，大庆油田开始了利用二氧化碳驱油的探索试验。

2022 年，中国石油二氧化碳（CO_2）年注入量达到 111 万吨，累计注入量达到 563 万吨。

In 1965, Daqing Oilfield began exploring and testing the use of carbon dioxide for oil displacement.

In 2022, CNPC's annual CO_2 injection amount reached 1.11 million tons, and the cumulative injection amount reached 5.63 million tons.

中国石油 CCUS 的发展总体上分为三个时段：

探索阶段（1965—2006 年）、攻关试验阶段（2007—2021 年）与工业示范阶段（2022 年及以后）。

The CCUS development of CNPC is divided into three stages: Exploration (1965-2006), Technical Research and Test (2007-2021), and Industrial Demonstration (2022 and beyond).

探索阶段（1965—2006年）(1)

攻关试验阶段（2007—2021年）(2)

探索阶段的突破性进展和标志性成果：

1965年
大庆油田碳酸水注入试验拉开我国探索 CO_2 驱油序幕。

1999年
吉林油田开展 CO_2 驱油先导试验。

2006年
首次提出中国发展CCUS/CCS产业倡议（香山科学会议）。

攻关试验阶段突破性进展和标志性成果：

2007年
启动国家973计划项目"温室气体提高石油采收率的资源化利用及地下埋存"。

2008年
中国石油承担国家科技重大专项"大型油气田及煤层气开发"下设项目及示范工程。

2009年
承担国家863计划项目"CO_2驱油提高石油采收率与封存关键技术研究"。

2009年
设立中国石油重大专项"吉林油田CO_2驱油与埋存关键技术研究"。

2013年
宁夏石化建成15万吨/年烟气CO_2捕集装置。

工业示范阶段（2022年及以后）(3)

2014年
吉林油田建成10万吨级CCUS-EOR全流程示范工程。

2019年
新疆准噶尔盆地CCUS项目（HUB）成为OGCI全球首批5个CCUS产业促进中心之一。

2021年
设立中国石油重大科技专项"二氧化碳规模化捕集、驱油与埋存全产业链关键技术研究及示范"。

2021年
成立中国石油二氧化碳捕集、利用与封存重点实验室和碳中和技术研发中心。

2021年
部署"四大工程示范"和"六个先导试验"CCUS工程。

工业示范阶段突破性进展和标志性成果：

2022年
提高油气采收率全国重点实验室获批建设。

2022年
启动建设松辽盆地300万吨CCUS示范工程。

2022年
发布《中国石油绿色低碳发展行动计划3.0》，确立CCUS发展战略。

2022年
长庆油田开启鄂尔多斯盆地千万吨级CCUS/CCS产业基地建设新征程。

Milestones in the first stage:

1965
The carbonated water injection test in Daqing oilfield opened the prologue of CO_2 flooding in China.

1999
A CO_2 flooding pilot test was carried out in Jilin Oilfield.

2006
China's initiative to develop CCUS/CCS industry was proposed at the Xiangshan Science Conference.

Milestones in the second stage:

2007
The National 973 Program "Resource Utilization and Underground Storage of Greenhouse Gas Enhanced Oil Recovery" was started.

2008
The National Science and Technology Major Project "Safety Development and Utilization Technology of CO_2 Containing Natural Gas Reservoir" was funded.

2009
CNPC undertook "The Study on Key Technologies of CO_2 Flooding for EOR and Storage", a project under the National 863 Program.

2009
The CNPC Major Project "Research on Key Technology of CO_2 Enhanced Oil Recovery and CO_2 Storage in Jilin Oilfield" was set up.

2013
Low CO_2 concentration flue gas capture equipment with a 1.5×10^5 tons/year processing scale in Ningxia PetroChemical Company was launched.

Industrial Demonstration Stage (2022 and beyond) (3)

2014
Jilin Oilfield completed a 1.0×10^5 tons CCUS-EOR life-cycle demonstration project.

2019
Xinjiang Hub became one of the first five OGCI CCUS industry promotion centers in the world.

2021
The CNPC Major Science and Technology Project "R&D on Key Technologies of the Whole Industrial Chain of Large-Scale CO_2 Capture, Oil Displacement, and Storage" was set up.

2021
The CNPC Key Laboratory of CO_2 Capture, Utilization, and Storage and the Carbon Neutral Technology R&D Center were established.

2021
CNPC deployed CCUS industrial construction of four demonstration projects and six pilot tests.

Milestones in the third stage:

2022
The National Key Laboratory of Enhanced Oil and Gas Recovery was approved for construction.

2022
A large-scale CCUS application project of 3 million tons in the Songliao Basin was launched.

2022
CNPC Green and Low-Carbon Development Action Plan 3.0 was released to decide on the next strategic objective of CNPC with CCUS as one of the emphasis.

2022
Changqing Oilfield started a new stage for constructing a 10 million tons CCUS/CCS industrial base in Ordos Basin.

2 中国石油 CCUS 大事记

2 Chronicle of Major Events on CCUS in CNPC

从 1965 年 9 月开始，中国石油 CCUS 事业发展历程三个阶段中的系列战略性举措、突破性发展、标志性成果共有 38 项大事。

Since September 1965, CNPC has fulfilled a total of 38 major events, including strategic measures, breakthroughs and landmark achievements during the three stages of CCUS business development.

2.1 探索阶段（1965—2006年）

2.1 1965—2006: Exploration Stage

1965年 大庆油田碳酸水注入试验拉开中国探索CO_2驱油序幕[①]

1965: Carbonated water injection test in Daqing Oilfield ushered in a new era of CO_2 flooding exploration in China[①]

1965年9月，大庆油田在萨尔图油田北2区东部葡Ⅰ4-7层开展2个井组14口井含3.8% CO_2碳酸水的小井距提高原油采收率试验（图2.1）。连续4个月注入CO_2总量为594吨，原油采收率比纯水驱井组提高4.4%。

In September 1965, Daqing Oilfield carried out an EOR test with 3.8% CO_2 carbonated water in 14 wells of 2 well groups in Pu I 4-7 layers with close spacing in the eastern part of the North 2 area of Saertu Oilfield (Fig. 2.1). The total amount of CO_2 injected for four consecutive months was 594 tons, and the crude oil recovery rate was 4.4% higher than that of the pure water flooding group.

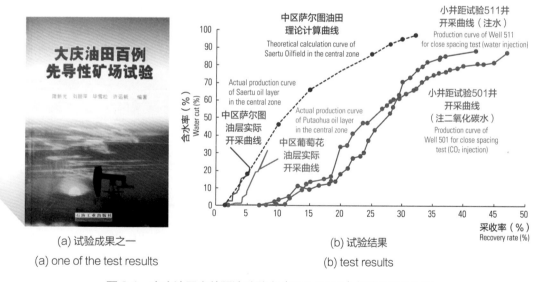

(a) 试验成果之一
(a) one of the test results

(b) 试验结果
(b) test results

图2.1 大庆油田小井距注碳酸水（CO_2 3.8%）提高采收率试验
Fig. 2.1 EOR Test with 3.8% CO_2 carbonated water injection at a close spacing in Daqing Oilfield

---

[①] 由大庆油田提供 (provided by Daqing Oilfield)。

1984年 大庆油田与法国公司合作开展 CO_2 驱油可行性及矿场试验研究[1]

1984: Daqing Oilfield partnered with French companies to carry out CO_2 flooding feasibility study and field pilot test [1]

1984年6月，大庆油田与法国德希尼布公司、法国石油研究院等签署技术服务合同，在萨南东部过渡带葡 I 2 层开展了4注9采 CO_2 驱油可行性研究与矿场先导试验，合同有效期截止到1993年6月。合作内容包括油藏研究、实验室评价、干扰试井方案设计、技术经济初步分析与矿场试验。项目期内累计注入 CO_2 总量为3.34万吨，累计产油1.43万吨，平均提高采收率4.67%（图2.2）。

In June 1984, Daqing Oilfield signed a technical service contract with France's Techinip Energies and IFP (Institut Francais du Pétrole) Energies Nouvelles to carry out CO_2 flooding feasibility study and field pilot test using 4 injection wells and 9 production wells in the Pu I 2 layer in the eastern Sanan transition zone, which was valid until June 1993. The content of the cooperation included reservoir research, laboratory evaluation, interference well test scheme design, technical and economic preliminary analysis, and field test. During the project period, 33,400 tons of CO_2 were injected and 14,300 tons of oil were produced, with an average EOR of 4.67% (Fig. 2.2).

图2.2　大庆油田与法国公司等开展 CO_2 驱油研究

Fig. 2.2　CO_2 flooding study between Daqing Oilfield and French companies

---

[1] 由大庆油田提供 (provided by Daqing Oilfield)。

1994年 吉林油田开展油田多井组 CO_2 吞吐试验

1994: Jilin Oilfield carried out a multi-well group CO_2 huff-and-puff test in the oilfield

1994 年，吉林新立油田开展 CO_2 吞吐增产先导试验，探索了弱驱替、水敏储层油井 CO_2 吞吐增产、CO_2 泡沫压裂技术的有效性。实施 224 口油井 CO_2 吞吐试验，其中有效井 179 口，有效率为 79.9%。累计增产原油 2 万吨，平均单井增产原油 89.4 吨。

In 1994, Jilin Xinli Oilfield carried out the pilot test of CO_2 huff-and-puff stimulation to explore the effectiveness of weak displacement, CO_2 huff-and-puff stimulation in the water-sensitive reservoir, and CO_2 foam fracturing technology. CO_2 huff-and-puff tests were carried out in 224 oil wells, of which 179 were effective, with an effective rate of 79.9%. The cumulative increase in crude oil production was 20,000 tons, with an average increase of 89.4 tons per well.

1999年 吉林油田开展 CO_2 驱油试注实验[1]

1999: Jilin Oilfield carried out CO_2 flooding trial injection experiment[1]

1999 年，吉林油田在新 228 区块开展了 CO_2 驱油试注试验，6 年作业期内共进行了 2 口井 3 井次的现场注入试验，累计增油 0.654 万吨。依托试验成果 2003 年制定了中华人民共和国石油天然气行业标准《油水井注二氧化碳安全技术要求》(SY/T 6565—2003)，明确了陆上油气田注 CO_2 的设计、施工、运行操作、安全管理等方面的安全要求。

In 1999, Jilin Oilfield carried out CO_2 flooding trial injection experiment in Xin 228 block. During the 6-year operation period, a total of 3 times of field injection tests in 2 wells were carried out, with a cumulative oil increase of 6,540 tons. Based on the test results, the People's Republic of China Standard of Petroleum and Natural Gas Industry *Technique Requirements for CO_2 Injection in Petroleum or Water Well Safety* (SY/T 6565—2003) was released in 2003, which defined the safety requirements in the design, construction, operation and safety management of CO_2 injection in onshore oil and gas fields.

---

[1] 由吉林油田提供 (provided by Jilin Oilfield)。

2003年　大庆油田开展特低渗透储层 CO_2 驱油先导性试验[1]

2003: Daqing Oilfield carried out a CO_2 flooding pilot test in ultra-low permeability reservoirs[1]

2003年3月，大庆油田在宋芳屯油田芳48断块扶杨油层开展了1注5采 CO_2 驱油先导性矿场试验。扶杨油层属于渗透率仅1毫达西左右的特低渗透储层，驱油技术实施的难度较大。五年内注入 CO_2 总量为2.57万吨，累计产油1.28万吨，通过 CO_2 驱油实现了特低渗透储层的有效动用（图2.3）。

In March 2003, Daqing Oilfield carried out a pilot field test of CO_2 flooding using l injection well and 5 production wells in Fuyang oil layer of Fang 48 fault block in Songfangtun Oilfield. Fuyang oil layer is an ultra-low permeability reservoir whose permeability is only about 1 millidarcy, so it is difficult to implement oil flooding technology. Within five years, the total amount of CO_2 injected was 25,700 tons, and the cumulative oil production was 12,800 tons, indicating that Daqing Oilfield achieved effective production of ultra-low permeability reservoir by CO_2 flooding (Fig. 2.3).

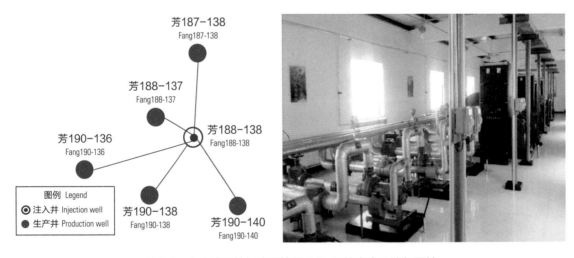

图 2.3　大庆油田特低渗透储层 CO_2 驱油方案及注气泵站

Fig. 2.3　CO_2 flooding scheme in ultra-low permeability reservoir and gas injection pumping station in Daqing Oilfield

---

[1] 由大庆油田提供 (provided by Daqing Oilfield)。

2004 年 中国寰球工程有限公司建成具有自主知识产权的 30 万吨/年中浓度 CO_2 低温甲醇洗分离技术[1]

2004: China Huanqiu Contracting & Engineering Co., Ltd. (HQC) built a 300,000 t/y medium-concentration CO_2 low-temperature methanol washing separation technology with independent intellectual property rights[1]

2004 年，中国寰球工程有限公司开发了国内首套具有自主知识产权的低温甲醇洗技术，获得授权国家发明专利一项（专利号 ZL 2005 1 0073353.5）。该技术吸收酸性气体时选择性高，CO_2 产品纯度大于 99%（体积分数），净化气中总硫含量不大于 0.1×10^{-6}（体积分数）。该技术于 2005 年成功应用于山东华鲁恒升化工股份有限公司 30 万吨/年煤基化肥项目，2008 年荣获国家科学技术进步奖二等奖（图 2.4）。

In 2004, HQC developed the first cryogenic methanol washing technology with independent intellectual property rights in China, and was granted a national invention patent (patent number ZL 2005 1 0073353.5). The technology has high selectivity when absorbing acid gas, with CO_2 product purity > 99vol% and purified gas total sulfur ≤ 0.1ppm. It was successfully applied to SHANDONG HUALU-HENGSHENG CHEMICAL CO., LTD 300,000 t/y coal-based fertilizer project in 2005, and won the second prize of National Science and Technology Progress Award in 2008 (Fig.2.4).

图 2.4　中国寰球工程有限公司低温甲醇洗国家科技进步奖证书及发明专利

Fig.2.4　National Science and Technology Progress Award Certificate for cryogenic methanol washing and invention patent of HQC

---

[1] 由中国寰球工程有限公司提供 (provided by HQC)。

2006年 中国石油在香山科学会议首次提出中国发展 CCUS/CCS 产业倡议[1]

2006: CNPC raised the initiative of developing the CCUS/CCS industry in China for the first time at the Xiangshan Science Conference [1]

2006年香山科学会议第276、第279次学术讨论会在中国北京香山召开，会议主题为温室气体地下封存及其在提高石油采收率中的资源化利用。中国石油专家首次提出了CO_2捕获、利用与封存技术（CCUS）概念，并提出了中国发展CCUS/CCS产业倡议。此次会议达成了两点共识：一是CO_2减排与利用必须紧密结合；二是CO_2主要利用途径是提高石油采收率（EOR）。

The 276th and 279th Symposiums of the 2006 Xiangshan Science Conference were held in Xiangshan, Beijing, China, with the theme of underground storage of greenhouse gases and their resource utilization in enhanced oil recovery. CNPC experts put forward the concept of CO_2 capture, utilization and storage (CCUS) technology for the first time, and raised the initiative of developing the CCUS/CCS industry in China. Two points of consensus were reached at the meeting: one is that CO_2 emission reduction and utilization must be closely integrated, and the other is that the main way of CO_2 utilization is enhanced oil recovery (EOR).

---

①由中国石油勘探开发研究院提供 (provided by RIPED)。

2.2 攻关试验阶段（2007—2021年）

2.2 2007-2021: Technical Research and Test Stage

2007年 中国石油承担国家973计划项目"温室气体提高石油采收率的资源化利用及地下埋存"[①]

2007: CNPC undertook the Utilization of Greenhouse Gas as Resource in EOR and Storage It Underground, a project under the National 973 Program [①]

2007年，国家科技部批准了我国第一个与CCUS技术相关的重大基础研究项目（973计划）"温室气体提高石油采收率的资源化利用及地下埋存"。中国石油开展了我国主要油区CO_2地质埋存量和驱油利用的潜力评价，初步形成了适合中国陆相沉积储层特点的CO_2高效利用和埋存技术体系，建成了国内首个集脱碳、CO_2地面集输、驱油与埋存等技术为一体的工业规模的试验系统（图2.5）。

In 2007, China's Ministry of Science and Technology approved the Utilization of Greenhouse Gases as Resource in EOR and Underground Storage, the country's first project related to CCUS technology under the National Basic Research Program (973 Program). CNPC carried out the potential evaluation of CO_2 geological storage and flooding utilization in China's main oil areas, preliminarily established an efficient CO_2 utilization and storage technology system suitable for the characteristics of China's continental sedimentary reservoirs, and built the first industrial-scale test system integrating decarbonization, CO_2 ground gathering and transportation, oil flooding and storage technologies in China (Fig. 2.5).

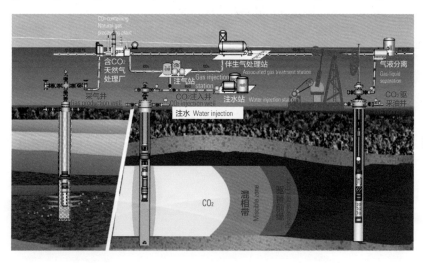

图2.5 含CO_2火山岩气藏开发及CO_2-EOR与埋存示意图

Fig. 2.5 Schematic diagram of CO_2-containing volcanic gas reservoir development, CO_2-EOR and storage

---

①由中国石油勘探开发研究院提供 (provided by RIPED)。

2008年 中国石油承担国家科技重大专项"大型油气田及煤层气开发"下设项目和示范工程[①]

2008: CNPC undertook a project under "Development of Large Oil and Gas Fields and Coalbed Methane", a national science and technology major project and demonstration [①]

2008年6月,中国石油承担国家科技重大专项"大型油气田及煤层气开发"下设项目和示范工程"含CO_2天然气藏安全开发与CO_2利用技术"。项目围绕含CO_2天然气藏开发和产出CO_2驱油等关键问题,初步形成了含CO_2天然气藏安全开发的应用基础理论,建立了CO_2提高低渗透油田采收率理论,优化了CO_2驱油设计方案(图2.6)。

In June 2008, CNPC undertook a project under the national science and technology major project "Development of Large Oil and Gas Fields and Coalbed Methane", and the demonstration project "Safety Development of CO_2-containing Natural Gas Reservoirs and CO_2 Utilization Technology". Focusing on key issues such as the development of CO_2-containing natural gas reservoirs and flooding with CO_2 produced, the project preliminarily formed the basic application theory for the safety development of CO_2-containing natural gas reservoirs, established the theory of CO_2-EOR in low-permeability oilfields, and optimized the CO_2 flooding design scheme (Fig. 2.6).

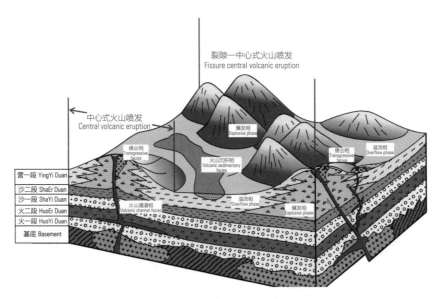

图 2.6 含CO_2天然气藏火山岩相概念模式

Fig. 2.6 Conceptual model of volcanic lithofacies in CO_2-containing natural gas reservoirs

---

① 由中国石油勘探开发研究院提供 (provided by RIPED)。

2009 年 中国石油承担国家 863 计划项目 "CO_2 驱油提高石油采收率与封存关键技术研究"[1]

2009: CNPC undertook the Study on Key Technologies of CO_2 Flooding for EOR and Storage, a project under the National 863 Program[1]

2009 年 6 月，中国石油承担国家 863 计划项目"CO_2 驱油提高石油采收率与封存关键技术研究"。针对 CO_2 驱油提高石油采收率与封存关键技术问题，建立了 CO_2 封存与提高采收率潜力评价方法，完成了 117 亿吨石油资源的 CO_2 提高采收率与封存潜力评价，形成五参数法 CO_2 分层测试及调配工艺并应用，并制定了 CO_2 驱油与封存技术经济评价方法标准和规范（图 2.7）。

In June 2009, CNPC undertook the Study on Key Technologies of CO_2 Flooding for EOR and Storage, a project under the National 863 Program. Aiming at key technical problems of CO_2 flooding for EOR and storage, a CO_2 storage and EOR potential evaluation method was established. The evaluation of CO_2-EOR and storage potential of 11.7 billion tons of petroleum resources was completed, the five-parameter CO_2 layered testing and blending process was established and applied, and the standards and specifications for the economic evaluation method of CO_2 flooding and storage technology were formulated (Fig. 2.7).

图 2.7　长岭气田脱碳装置及 CO_2 驱产出气超临界注入装置

Fig. 2.7　Decarbonization plant and CO_2 flooding produced gas supercritical injection plant in Changling Gas Field

---

[1] 由中国石油勘探开发研究院提供 (provided by RIPED)。

2009年 中国石油重大科技专项"吉林油田 CO_2 驱油与埋存关键技术研究"启动[1]

2009: CNPC launched a major project "Study on Key Technologies of CO_2 Flooding and Storage in Jilin Oilfield"[1]

2009年7月,中国石油启动重大科技专项"吉林油田 CO_2 驱油与埋存关键技术研究"。该项目开发低渗透油藏 CO_2 混相驱提高采收率与 CO_2 埋存技术,建立 CO_2 驱油和埋存先导试验基地。项目期内共试验黑59、黑79南两个区块,黑59区块累计注入27.3万吨,相较于水驱,CO_2 混相驱日产油量最大提高69%,采收率提高约10.4%;黑79区块南累计注入48.5万吨,日产油量最大提高33%,提高采收率约14.5%(图2.8)。

In July 2009, CNPC launched a major project "Study on Key Technologies of CO_2 Flooding and Storage in Jilin Oilfield", aiming to develop CO_2 miscible flooding for EOR and CO_2 storage technologies in low-permeability reservoirs and establish a pilot test base for CO_2 flooding and storage. During the project period, a total of two blocks, Hei 59 and Hei 79 South, were tested. In Hei 59 block, the cumulative injection was 273,000 tons. Compared with water flooding, the daily oil production after CO_2 miscible flooding increased by up to 69%, and the oil recovery rate rose by about 10.4%. In Hei 79 South block, the cumulative injection was 485,000 tons, the daily oil production increased by up to 33%, and the oil recovery rate rose by about 14.5% (Fig. 2.8).

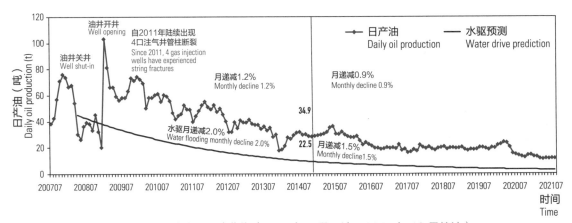

图2.8 黑59试验区采油曲线(2008年5月开注、2014年10月停注)

Fig. 2.8 Oil production curve of Hei 59 test area
(injection began in May 2008 and stopped in October 2014)

---

[1]由吉林油田提供 (provided by Jilin Oilfield)。

2011年 设立"中国石油低碳关键技术研究"重大科技专项进行 CO_2 捕集与封存技术攻关和示范[①]

2011: CNPC set up a major science and technology project "Research on Key Low-Carbon Technologies of CNPC" to explore and demonstrate CO_2 capture and storage technologies[①]

2011年4月,"中国石油低碳关键技术研究"重大科技专项启动。针对低浓度 CO_2 捕集成本较高的技术挑战,开发的三元溶剂 CO_2 循环吸收容量较常用的单乙醇胺溶液提高了30%,捕集工艺能耗降低30%,总成本降低40元/吨 CO_2。综合考虑 CO_2 在咸水层封存的动态变化及主要形式,提出了适合我国 CO_2 咸水层封存潜力评价方法,明确新疆等4大油田83个咸水层构造的有效 CO_2 封存潜力约270亿吨(图2.9)。

In April 2011, CNPC launched a major science and technology project "Research on Key Low-Carbon Technologies of CNPC". To meet the technical challenges of the high cost of capturing low-concentration CO_2, the CO_2 circular absorption capacity of developed ternary solvent was increased by 30% compared with the commonly used monoethanolamine solution, the energy consumption of the capture process was reduced by 30%, and the total cost was decreased by RMB 40 per ton of CO_2. Considering the dynamic changes and main forms of CO_2 stored in saline aquifers, a method suitable for evaluating the CO_2 storage potential in saline aquifers in China was proposed, and the effective CO_2 storage potential of 83 saline aquifers in 4 major oilfields including Xinjiang Oilfield was identified at about 27 billion tons (Fig. 2.9).

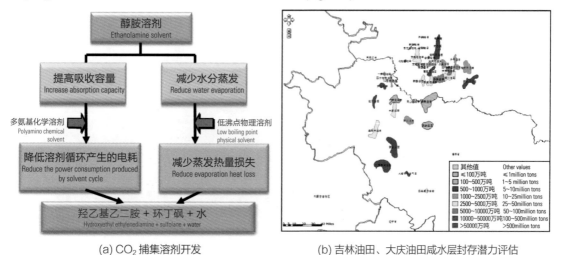

(a) CO_2 捕集溶剂开发 (b) 吉林油田、大庆油田咸水层封存潜力评估
(a) Development of CO_2 trapping solvent (b) storage potential evaluation in saline aquifers in Jilin and Daqing Oilfields

图 2.9 低碳关键技术研究
Fig. 2.9 Research on key low-carbon technologies

[①]由安全环保技术研究院提供 (provided by RISE)。

2013年 宁夏石化公司建成 15 万吨/年低浓度烟气 CO_2 捕集装置[①]

2013: Ningxia PetroChemical Company built a 150,000 t/y CO_2 capture plant from low-concentration flue gas [①]

2013 年，宁夏石化公司建成 15 万吨/年转化炉烟道气 CO_2 捕集装置，其中烟道气中 CO_2 浓度为 7.31%。该装置由回收、压缩、干燥三部分组成，捕集后的 CO_2 送往尿素装置做原料气使用。CO_2 捕集过程采用 IST-aMEA 溶液作为吸收剂，吸收 CO_2 后的富液经加热再生后循环使用（图 2.10）。

In 2013, Ningxia PetroChemical Company built a 150,000 t/y CO_2 capture plant from low-concentration flue gas, with a CO_2 concentration in the flue gas of 7.31%. The plant consists of three parts: recovery, compression and drying, and the captured CO_2 is sent to the urea unit for feedstock gas. In the CO_2 capture process, IST-aMEA solution is used as the absorbent, and the flooded liquid after absorbing CO_2 could be recycled after heating and regeneration (Fig. 2.10).

图 2.10 宁夏石化公司 15 万吨/年烟气 CO_2 捕集装置图

Fig. 2.10 150,000 t/y CO_2 capture plant from low-concentration flue gas of Ningxia PetroChemical Company

---

[①]由宁夏石化公司提供 (provided by Ningxia PetroChemical Company)。

2013年 大庆油田建成 13.5 千米 CO_2 气态输送管道[1]

2013: Daqing Oilfield completed a 13.5km gaseous CO_2 transmission pipeline [1]

2013年10月,大庆油田建成自徐深9液化站至树101-CO_2液化注入站间的气态输送管道,用于将天然气分离装置回收的CO_2输送至驱油注入井组。管道全长13.5千米,直径是219毫米,设计压力为4兆帕。截至2022年10月末,累计输送CO_2总量为79.1万吨(图2.11)。

In October 2013, Daqing Oilfield completed a gaseous CO_2 transmission pipeline from Xushen 9 Liquefaction Station to Shu 101 CO_2 Liquefaction Injection Station to transport CO_2 recovered from the natural gas separation unit to the flooding injection well group. The total length of the pipeline is 13.5 km, the diameter is 219 mm, and the design pressure is 4 MPa. By the end of October 2022, the total amount of CO_2 transported was 791,000 tons (Fig. 2.11).

(a) 大庆油田气态输送管道示意图

(a) Schematic diagram of gaseous CO_2 transmission pipeline in Daqing Oilfield

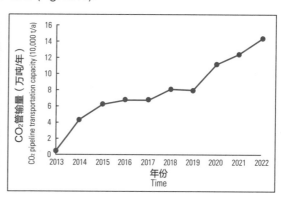

(b) 历年输送量

(b) transportation volume over the years

图 2.11 大庆油田 CO_2 气态输送管道

Fig. 2.11 Gaseous CO_2 transmission pipeline in Daqing Oilfield

---

[1] 由大庆油田提供 (provided by Daqing Oilfield)。

2014年 吉林油田建成十万吨级 CCUS-EOR 全流程示范工程[①]

2014: Jilin Oilfield completed a 100,000-ton full-process CCUS-EOR demonstration project [①]

2014年，吉林油田建成包括捕集、输送、注入采出、集输处理及循环的十万吨级全流程 CCUS-EOR 工业化示范工程。6年内累计注入 CO_2 总量为80万吨，累计增产原油8万吨（图2.12）。

In 2014, Jilin Oilfield completed a 100,000-ton full-process CCUS-EOR industrialization demonstration project including capture, transportation, injection and production, gathering and transportation treatment, and circulation. Within six years, the total amount of CO_2 injection was 8×10^5 tons, and the cumulative increase in crude oil production was 80,000 tons (Fig. 2.12).

图 2.12 吉林油田全流程 CCUS 项目

Fig. 2.12 Photo of full-process CCUS-EOR demonstration project in Jilin Oilfield

---

①由吉林油田提供 (provided by Jilin Oilfield)。

2014年 大庆油田开展特低渗透油层 CO_2 非混相驱油工业化示范应用试验[①]

2014: Daqing Oilfield carried out an industrial application test of CO_2 non-miscible flooding in an ultra-low permeability reservoir [①]

2014年9月,大庆油田在榆树林油田树101、树16区块开展了十万吨级 CO_2 非混相驱现场试验,形成了 CO_2 捕集、管输、液化、注入与埋存全流程技术体系。截至2022年10月,累计注入 CO_2 总量为116万吨,产油38.8万吨,原油采收率提高约9%(图2.13)。

In September 2014, Daqing Oilfield carried out a 100,000-ton CO_2 non-miscible flooding field test in Shu 101 and Shu 16 blocks of Yushulin Oilfield, forming a full-process technical system of CO_2 capture, pipeline transportation, liquefaction, injection and storage. As of October 2022, the total amount of CO_2 injection was 1.16 million tons, oil production was 388,000 tons, and the crude oil recovery rate increased by about 9% (Fig. 2.13).

(a) CO_2 液化系统　　　　　　　　　　(b) CO_2 压缩注入系统
(a) CO_2 Liquefaction system　　　　　(b) CO_2 compression injection system

图 2.13　大庆油田 CO_2 液化系统及压缩注入系统

Fig. 2.13　CO_2 liquefaction system and compression injection system in Daqing Oilfield

---

[①] 由大庆油田提供 (provided by Daqing Oilfield)。

2017年　中国石油参与 CCUS 国际标准（ISO/TR 27915）制定[1]

2017: CNPC participated in the formulation of the CCUS international standard (ISO/TR 27915) [1]

2017年8月，中国石油参与完成的《二氧化碳捕集、运输与地质封存——量化与核查》(Carbon dioxide capture, transportation, and geological storage — Quantification and Verification)（ISO/TR 27915:2017）正式发布，为 CCUS 项目减排量核查核证国际规则制定奠定了基础。多年来，中国石油积极参与国际标准化组织二氧化碳捕集运输与封存技术委员会（ISO/TC 265）工作，8 位中国石油专家成长为 TC 265 注册专家。

In August 2017, *Carbon Dioxide Capture, Transportation, and Geological Storage — Quantification and Verification* (ISO/TR 27915:2017), completed with the participation of CNPC, was officially released, which laid the foundation for the formulation of international rules for verification and certification of emission reductions of CCUS projects. Over the years, CNPC has actively participated in the work of the Technical Committee ISO/TC 265, Carbon Dioxide Capture, Transportation, and Geological Storage, and eight experts of ISO have become registered experts of the committee.

---

[1] 由安全环保技术研究院提供 (provided by RISE)。

2017年　南方石油勘探开发有限责任公司启动 CO_2 提高油气采收率与埋存先导试验[①]

2017: China Southern Petroleum Exploration & Development Corporation launched the pilot test of CO_2-enhanced oil and gas recovery and storage [①]

2017年3月，南方石油勘探开发有限责任公司启动福山油田 CO_2 提高油气采收率和埋存先导试验一期示范工程，将采出的富含 CO_2 天然气管输至花场处理站，经分离液化后用槽车拉运至注气现场（图2.14）。项目期内累计注入 CO_2 超15.8万吨，增油1.4万吨。二期示范工程在福山油田花场处理中心建成一套10万吨/年 CO_2 捕集、液化装置，形成自主的 CO_2 注采、集输、储运、处理技术体系，并于2021年10月投产试运，CO_2 注入能力达20万吨/年（图2.15）。

In March 2017, China Southern Petroleum Exploration & Development Corporation launched the first-phase demonstration pilot test of CO_2-enhanced oil and gas recovery and storage in Fushan Oilfield. The produced CO_2-rich natural gas was piped to the Huachang treatment station, and was then transported to the gas injection site by tanker after separation and liquefaction (Fig. 2.14). During the project period, more than 158,000 tons of CO_2 were injected, and oil production increased by 14,000 tons. In the second-phase demonstration project, a 100,000 t/y CO_2 capture and liquefaction plant was built in Huachang Treatment Center of Fushan Oilfield, forming an independent technical system of CO_2 injection and production, gathering and transportation, storage and transportation, and treatment. The plant was put into trial operation in October 2021, with a CO_2 injection capacity of 200,000 t/y (Fig. 2.15).

图 2.14　福山油田 CCUS 先导试验一期装置

Fig. 2.14　Phase I plant of CCUS pilot test in Fushan Oilfield

图 2.15　福山油田 CCUS 先导试验二期装置

Fig. 2.15　Phase II plant of CCUS pilot test in Fushan Oilfield

---

[①] 由南方石油勘探开发有限责任公司提供 (provided by China Southern Petroleum Exploration & Development Corporation)。

2018年　中国石油参加巴西 CCUS 项目[1]

2018: CNPC participated in a CCUS project in Brazil [1]

2018年1月,中国石油参加的巴西项目先锋号FPSO(海上浮式生产储油船)开始进行CO_2回注。该项目原油处理能力5万桶/天,气处理能力400万立方米/天。采用膜分离法回收采出气中的高浓度CO_2(44%),并加压回注地层(图2.16)。

In January 2018, the Pioneero de Libra FPSO (offshore floating production and storage tanker) in Brazil, in which CNPC participated, began CO_2 reinjection. It had a crude oil processing capacity of 50,000 bbl/d and a gas processing capacity of 4 million m³/d. The high-concentration CO_2 (44%) in the produced gas was recovered by membrane separation, and then pressurized and reinjected into the formation (Fig. 2.16).

图 2.16　巴西里贝拉项目先锋号 FPSO

Fig. 2.16　Pioneero de Libra FPSO in Brazil

---

[1] 由中国石油国际勘探开发有限公司提供 (provided by CNODC)。

2019年　新疆油田启动 CO_2 混相驱先导试验[1]

2019: Xinjiang Oilfield started the pilot test of CO_2 miscible flooding [1]

2019年1月，新疆油田启动了八区530井区东部克下组砾岩油藏15注43采 CO_2 混相驱先导试验，探索低渗透水敏砾岩油藏 CO_2 驱提高采收率的可行性。覆盖地质储量约200万吨，年注入10万吨 CO_2，提高原油采收率25%以上（图2.17）。

In January 2019, Xinjiang Oilfield started the pilot test of CO_2 miscible flooding with the setting of 15 injection and 43 production wells at Kexia Formation in the east of Well Area 530, Area 8. The objective was to explore the feasibility of CO_2 flooding in low-permeability water-sensitive conglomerate reservoirs. The geological reserves were considered about 2 million tons. The annual injection amount was 100,000 tons of CO_2 and the oil recovery increased by more than 25% (Fig. 2.17).

图 2.17　新疆油田 CO_2 混相驱先导试验区

Fig. 2.17　Pilot test area of CO_2 miscible flooding in Xinjiang Oilfield

---

[1] 由新疆油田提供 (provided by Xinjiang Oilfield)。

2019年　首次在中国举办 OGCI 系列工作会议[1]

2019: A series of OGCI working meetings held firstly in China [1]

2019年6月，中国石油在杭州承办了油气行业气候倡议组织（OGCI）执委会、CCUS工作组会议，并主办了"OGCI中国碳捕集封存与利用商业化机遇论坛"。来自国家科技部、生态环境部、国内油气企业及高校的近150位专家学者、OGCI成员公司与气候基金约50名外方代表围绕CCUS技术商业化问题开展交流探讨。本次会议是第一次在中国召开，会议就《OGCI CCUS 撬动者宣言》的内容基本达成一致，推动了OGCI成员公司在全球布局CCUS五大区域中心。

In June 2019, a series of working meetings of the Oil and Gas Climate Initiative (OGCI), organized by CNPC, were held in Hangzhou. We also hosted the OGCI China CCUS Commercialization Opportunities Forum. Nearly 150 representatives from the Ministry of Science and Technology, and the Ministry of Ecology and Environment, domestic oil and gas companies, universities, and about 50 foreign representatives from OGCI member companies and the Climate Fund attended the meetings, which were firstly held in China. In-depth discussions on the commercialization of CCUS technology were conducted, and a consensus was reached regarding *CCUS Kickstarter Initiative, OGCI*, advancing the global layout of five CCUS hubs in the world by OGCI member companies.

1 由安全环保技术研究院提供 (provided by RISE).

2019年　新疆 HUB 成为 OGCI 全球首批 5 个 CCUS 产业促进中心之一[1]

2019: Xinjiang Hub selected as one of OGCI's first five CCUS hubs in the world [1]

2019 年 9 月 22—23 日，油气行业气候倡议组织（英文简称 OGCI）2019 年度企业领导人峰会在美国纽约召开，会上正式宣布将实施"碳捕集、利用与封存（CCUS）撬动者计划"。由中国石油牵头的新疆准噶尔项目与挪威石油公司牵头的挪威北极光项目、壳牌公司牵头的荷兰鹿特丹项目、美国雪佛龙公司和西方石油公司联合牵头的美国墨西哥湾项目、英国石油公司牵头的英国提赛德项目共同成为 OGCI 全球首批布局的 5 个 CCUS 产业促进中心（图 2.18）。

On September 22-23, 2019, the OGCI Business Leadership Summit 2019 was held in New York, officially announcing the implementation of CCUS Kickstarter Initiative. The Xinjiang Hub led by CNPC was selected as one of OGCI's first five CCUS hubs in the world. Other hubs included the Northern Lights led by Statoil ASA, the Rotterdam project led by Shell, the Gulf of Mexico project jointly led by Chevron and OXY, and the Net Zero Teesside Project in the UK led by BP (Fig. 2.18).

图 2.18　新疆 HUB CCUS 规划图

Fig. 2.18　Planning map of Xinjiang CCUS Hub

---

[1] 由新疆油田提供 (provided by Xinjiang Oilfield)。

2021年　成立中国石油二氧化碳捕集、利用与封存重点实验室[1]

2021: The Key Laboratory of Carbon Capture Utilization and Storage, CNPC established[1]

2021年8月30日，中国石油CCUS重点实验室正式挂牌运行，旨在推动二氧化碳捕集、驱油利用、地质封存等方面的基础理论研究与技术进步，重点开展CO_2捕集、CO_2地质及化工利用等研究，以期建成CO_2捕集、利用、封存和发展战略4大专业化实验室和1个CO_2驱油与埋存试验基地，成为CCUS超前基础理论研究和新技术新方法研发验证平台（图2.19）。

On August 30, 2021, the Key Laboratory of Carbon Capture Utilization and Storage, CNPC was officially put into operation. It aims to promote basic theoretical research and technological progress in carbon dioxide capture, oil flooding and geological storage, with an emphasis on CO_2 capture, CO_2 geological and chemical utilization. It has a vision to build four specialized sub-laboratories of CO_2 capture, utilization, storage and development strategy and a CO_2 flooding and storage test base, and to become a platform for the research of CCUS advanced basic theories and verification of new technologies and methods (Fig. 2.19).

图 2.19　中国石油二氧化碳捕集、利用与封存重点实验室

Fig. 2.19　Key Laboratory of Carbon Capture Utilization and Storage, CNPC

---

[1] 由中国石油勘探开发研究院提供 (provided by RIPED)。

| 2021年 | 中国石油重大科技专项"二氧化碳规模化捕集、驱油与埋存全产业链关键技术研究及示范"启动① |

2021: CNPC launched a major science and technology project "Research and Demonstration of Key Technologies for the Whole Industry Chain of Scalable Carbon Dioxide Capture, Oil Flooding and Storage" ①

2021年9月，中国石油重大科技专项"二氧化碳规模化捕集、驱油与埋存全产业链关键技术研究及示范"启动。该专项共设置"炼化企业低浓度二氧化碳低成本高效捕集技术研究"等8个课题，开展炼化、管输、驱油与埋存等CCUS全产业链环节的专项研究与攻关设计，为打造CCUS原创技术策源地，促进全产业链高质量发展目标提供支撑。

In September 2021, CNPC launched a major science and technology project "Research and Demonstration of Key Technologies for the Whole Industry Chain of Scalable Carbon Dioxide Capture, Oil Flooding and Storage". The project has set up a total of 8 topics, including "Research on Low-cost and Efficient Low-concentration CO_2 Capture Technology of Refining and Chemical Enterprises". Special research and design of the whole CCUS industry chain covering refining and chemical, pipeline transportation, oil flooding and storage have been carried out. With these efforts, CNPC aims to provide support for creating the source of CCUS original technology and achieving the goal of high-quality development of the whole industry chain.

① 由中国石油勘探开发研究院提供 (provided by RIPED)。

2021年 成立中国石油碳中和技术研发中心[1]

2021: R&D Center for Carbon Neutralization Technology established [1]

2021年12月,"中国石油天然气集团有限公司碳中和技术研发中心"获批成立。该中心围绕石油石化行业绿色低碳转型、碳达峰碳中和等重大技术需求,开展CCUS、甲烷控排、节能与流程再造等技术的基础理论、技术攻关与成果转化,为实现石油石化行业"双碳"目标提供持续技术支撑(图2.20)。

In December 2021, R&D Center for Carbon Neutralization Technology, CNPC was approved to be established. Focusing on the major technical needs of the petroleum and petrochemical industry, such as green and low-carbon transition, carbon peaking and carbon neutrality, the Center carried out basic theoretical and technical research and results transformation in areas of CCUS, methane emission control, energy saving and process reengineering, providing continuous technical support for achieving the "dual carbon" goals of the petroleum and petrochemical industry (Fig. 2.20).

图2.20 中国石油碳中和技术研发中心

Fig. 2.20 R&D Center for Carbon Neutralization Technology, CNPC

---

[1] 由安全环保技术研究院提供 (provided by RISE)。

2021年　中国石油担任全国 CCUS 标准化工作组组长单位[①]

2021: CNPC served as the leader of the National CCUS Standardization Working Group [①]

2021年3月，全国碳排放标准化技术委员会、全国能源基础与管理标准化委员会、全国环境管理标准化技术委员会联合发起并成立了全国 CCUS 标准化工作组，由中国标准化研究院、中国21世纪议程管理中心等单位的17名专家组成，中国石油作为组长单位，推荐专家担任组长，积极开展 CCUS 领域的标准体系及关键标准研究制定工作（图2.21）。

In March 2021, the National Carbon Emission Management Standardization Technical Committee, the National Technical Committee for Standardization of Energy Foundation and Management, and the National Technical Committee for Standardization of Environmental Management jointly initiated and established the National CCUS Standardization Working Group. Composed of 17 experts from the China National Institute of Standardization, the Administrative Center for China's Agenda 21, etc., with an expert from CNPC serving as the leader, the Working Group is committed to actively carrying out the research and formulation of the standard system and key standards in the field of CCUS (Fig. 2.21).

图 2.21　成立全国 CCUS 标准化工作组的通知

Fig.2.21　Notice of the Establishment of the National CCUS Standardization Working Group

---

[①] 由安全环保技术研究院提供 (provided by RISE).

2021年 牵头编写的《OGCI 中国 CCUS 商业化白皮书》正式发布[1]

2021: *OGCI CCUS in China Commercialization White Paper*, prepared under the leadership of CNPC, released officially [1]

2021年9月，由中国石油牵头编写的《OGCI 中国 CCUS 商业化白皮书》通过 OGCI 正式发布。白皮书着眼于发掘 CCUS 的价值和部署机会，探讨 CCUS 在创造气候、社会和经济效益的潜力，提高其在低碳能源发展、保障能源安全供给的市场地位，为应对气候变化、实现油气行业绿色低碳转型贡献中国石油力量（图 2.22）。

In September 2021, *OGCI CCUS in China Commercialization White Paper*, prepared under the leadership of CNPC, was officially released through OGCI. It focuses on exploring the value and deployment opportunities of CCUS, tapping the potential of CCUS to create climate, social and economic benefits, and improving CCUS's market position in low-carbon energy development and ensuring energy security & supply. It aims to contribute to the response to climate change and the green and low-carbon transition of the oil and gas industry (Fig. 2.22).

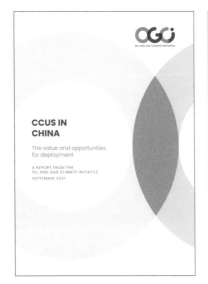

图 2.22 《OGCI 中国 CCUS 商业化白皮书》发布

Fig. 2.22 *OGCI CCUS in China Commercialization White Paper* released

---

[1] 由安全环保技术研究院提供 (provided by RISE)。

2021年 部署"四大工程示范"和"六个先导试验"CCUS工程,年注入能力超50万吨[1]

2021: The CCUS industrialization project including 4 demonstration projects and 6 pilot tests with an annual injection capacity of 500,000 tons was pushed forward [1]

2021年,中国石油部署吉林、大庆、长庆、新疆四个油田CCUS工业化示范工程项目,预计2025年CO_2年注入量355万吨,产油量110万吨。启动开展辽河、冀东、大港、华北、吐哈、南方等油田的碳驱油碳埋存六个先导试验,预计2025年CO_2年注入量81万吨,产油量31万吨。

In 2021, CNPC deployed four CCUS industrialization demonstration projects in Jilin, Daqing, Changqing and Xinjiang oilfields, with an estimated annual CO_2 injection of 3.55 million tons, and oil production of 1.10 million tons in 2025. It also launched six pilot tests on CO_2 flooding and sequestration in Liaohe Oilfield, Jidong Oilfield, Dagang Oilfield, Huabei Oilfield, Tuha Oilfield and Nanfang Oilfield, with an estimated annual CO_2 injection of 810,000 tons, and oil production of 310,000 tons in 2025.

---
[1] 由中国石油勘探开发研究院提供 (provided by RIPED)。

2021年 大港油田实施深层低渗油藏 CO_2 驱油埋存井组试验[1]

2021: The storage well group test of CO_2 flooding in deep and low-permeability reservoirs was carried out in Dagang Oilfield [1]

2021年4月，大港叶三拨油田CO_2混相驱井组试验开始运行。该试验采用2注5采井组水气交替注入，设计每周期CO_2注入量1.44万吨，水注入量6.6万立方米。截至2022年10月累计注入1.16万吨CO_2，阶段注水2.22万立方米，增产原油0.27万吨（图2.23）。

In April 2021, the well group test of CO_2 miscible flooding in Dagang Yesanbo Oilfield started to run. In this test, the water-gas alternate injection was adopted in a 2-injection and 5-production well groups, and the designed CO_2 injection volume and water injection volume per cycle were 14,400 tons and 66,000 m^3 respectively. By October 2022, a total of $1.16×10^4$ tons of CO_2 had been injected, 22,200 m^3 of water had been injected in stages, and the production of crude oil had increased by 2,700 tons (Fig. 2.23).

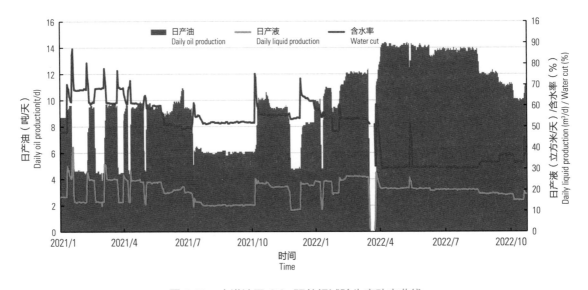

图 2.23 大港油田 CO_2 驱井组试验生产动态曲线

Fig. 2.23 Dynamic curve of experimental production of CO_2 flooding well group in Dagang Oilfield

---

[1] 由大港油田提供 (provided by Dagang Oilfield)。

2.3　工业示范阶段（2022 年及以后）

2.3　2022 and Beyond: Industrial Demonstration Stage

2022 年　中国石油召开 CCUS 工作推进会，启动建设松辽盆地 300 万吨 CCUS 示范工程[1]

2022: CNPC held a CCUS work promotion meeting, at which a 3-million-ton CCUS demonstration project in Songliao Basin launched [1]

为了全面落实中国石油绿色低碳发展战略和"三步走"总体部署，加快推动 CCUS 产业化发展，2022 年 2 月，组织召开中国石油 CCUS 重点工作推进会，积极推进 CCUS 重大科技专项研究，加快松辽盆地 300 万吨 CCUS 示范工程建设以及渤海湾、吐哈、南方先导试验，并抓好长庆、新疆 CCUS 工业应用布局，为推动实现"双碳"目标贡献石油力量（图 2.24）。

To fully implement its green and low-carbon development strategy and the "three-step" roadmap, and accelerate CCUS industrialization, CNPC convened a CCUS key work promotion meeting in February 2022. At the meeting, CNPC announced efforts to promote major CCUS science and technology research, advance the 3-million-ton CCUS demonstration in Songliao Basin and pilot tests in Bohai Bay, Tuha, and Nanfang oilfields, and enhance the layout of CCUS industrial applications in Changqing and Xinjiang oilfields, aiming to contribute to China's "dual carbon" goals (Fig. 2.24).

图 2.24　长庆油田二氧化碳驱橇装化布站效果及部分橇装设备

Fig. 2.24　Effect of CO_2 flooding skid mounted station layout and partial skid mounted equipment in Changqing Oilfield

---

[1] 由中国石油勘探开发研究院提供 (provided by RIPED)。

2022年　新疆油田制订 CCUS/CCS 双千万吨产业发展远景规划[1]

2022: Xinjiang Oilfield developed a long-term plan for the double ten-million-ton CCUS/CCS industry development [1]

2022年2月，新疆油田制订准噶尔盆地CCUS/CCS产业发展规划，计划2035年实现盆地1000万吨/年驱油、1000万吨/年咸水层埋存及盆地CO_2管输环网建设的目标。目前6个试验区已实现CO_2年注入量8.3万吨，2025年计划建成200万吨/年的CCUS产业示范工程基地。

In February 2022, Xinjiang Oilfield formulated a plan for the industrial development of CCUS/CCS in Junggar Basin, setting the goals of achieving 10 million t/y of CO_2 flooding, 10 million t/y of CO_2 storage in saline aquifers and construction of CO_2 ring pipeline by 2035. Up to now, the annual CO_2 injection volume of the six test areas has reached 83,000 tons. A 2 million t/y CCUS industrial demonstration base is planned to be built in 2025.

---

[1]由新疆油田提供 (provided by Xinjiang Oilfield)。

2022年 中国石油成立 CCUS 工作专班[1]

2022: CNPC established a CCUS working group [1]

为加快推动中国石油 CO_2 埋存与驱油工程产业化发展，全面推进百亿方专项工程建设，2022年4月15日，成立中国石油二氧化碳埋存与驱油工程工作专班（简称 CCUS 工作专班），确定了 CCUS 工作专班主要职责、组成人员，组建了 CCUS 专家支持团队，确定了办事机构和定期推进机制。

To speed up its pace towards the industrialization of CO_2 storage and flooding and promote the construction of its ten-billion cubic meters production capacity project, CNPC established a working group on CO_2 storage and flooding ("CCUS working group") on April 15, 2022. The main responsibilities and composition of the working group were identified, and an expert support team and relevant offices and regular promotion mechanisms were set up.

---

[1]由勘探板块提供 (provided by the exploration segment)。

2022年 提高油气采收率全国重点实验室获批建设[1]

2022: State Key Laboratory of Enhanced Oil Recovery approved for construction [1]

作为我国首批依托企业建设的实验室，提高石油采收率国家重点实验室于2010年正式挂牌运行。2022年，作为全国20家重点实验室之一，首批通过"提高油气采收率全国重点实验室"重组。实验室围绕油气系统形成与资源预测、油气储层与渗流、CCUS提高采收率等方向开展基础理论与应用技术研究，以期为国家"双碳"战略以及国家油气业务高质量发展发挥重要作用（图2.25）。

Among the first laboratories built by enterprises in China, the State Key Laboratory of Enhanced Oil Recovery was officially put into operation in 2010. In 2022, it was approved for restructuring as one of the 20 key laboratories in the country. The laboratory carries out basic theory and applied technology research on oil and gas system formation and resource forecasting, oil and gas reservoir and seepage, CCUS-EOR, etc., aiming to play an important role in China's "dual carbon" strategy and high-quality development of oil and gas industry (Fig. 2.25).

图 2.25 提高石油采收率国家重点实验室

Fig. 2.25 State Key Laboratory of Enhanced Oil Recovery

---

[1] 由中国石油勘探开发研究院提供 (provided by RIPED)。

2022年 《中国石油绿色低碳发展行动计划3.0》发布[①]

2022: *CNPC Green and Low-Carbon Development Action Plan 3.0* released [①]

2022年6月,《中国石油绿色低碳发展行动计划3.0》发布,制订"清洁替代、战略接替、绿色转型"的三步走总体部署,规划CCUS产业链建设工程、部署研究CCUS超前技术等三大方向,明确到2025年达370万吨/年、到2035年达2500万吨/年与到2050年达1亿吨/年的内部CO_2注入能力目标(图2.26)。

In June 2022, *CNPC Green and Low-Carbon Development Action Plan 3.0* was released. According to the plan, a three-step overall scheme of "clean substitution, strategic succession, and green transition" was established, three major directions such as the construction of the CCUS industrial chain were planned, CCUS advanced technology research were deployed and studied, etc. and the internal CO_2 injection capacity targets of 3.7 million t/y by 2025, 25 million t/y by 2035 and 100 million t/y by 2050 were defined (Fig. 2.26).

图 2.26 《中国石油绿色低碳发展行动计划3.0》发布

Fig. 2.26 *CNPC Green and Low-Carbon Development Action Plan 3.0* released

---

[①]由安全环保技术研究院提供 (provided by RISE)。

2022年 中国石油 CO_2 年注入量实现 111 万吨，累计注入量 563 万吨[1]

2022: CNPC achieved annual CO_2 injection of 1.11 million tons, with a cumulative injection volume of more than 5.63 million tons [1]

多年来，中国石油高度重视 CCUS 产业发展，推进二氧化碳提高原油采收率（CCUS-EOR）技术攻关，创新形成了 CCUS 全产业链技术体系，二氧化碳注入规模保持国内领先水平，推动我国 CCUS 项目在驱油利用领域迈入工业化示范应用阶段。截至 2022 年底，当年 CO_2 注入量 111 万吨（内部碳源占 51.6%）、增油 24.8 万吨，其中吉林油田注入 43 万吨、大庆油田注入 30 万吨，累计 CO_2 注入量超过 563 万吨（图 2.27）。

Over the years, CNPC has attached great importance to the industrial development of CCUS. It has stepped up research efforts in CCUS-EOR technology and innovatively established a CCUS whole industry chain technology system. It has kept a leading position in the scale of CO_2 injection and storage in China, and moved forward the country's CCUS projects towards industrial demonstration and applications in the field of CO_2 flooding. By the end of 2022, CNPC had injected 1.11 million tons of CO_2 (internal carbon sources accounted for 51.6%), with oil production increasing by 248,000 tons. Among them, Jilin Oilfield injected 430,000 tons and Daqing Oilfield injected 300,000 tons, with a cumulative CO_2 injection volume of more than 5.63 million tons (Fig. 2.27).

图 2.27 CO_2 高压超临界注入压缩机

Fig. 2.27 CO_2 high pressure supercritical injection compressor

---

[1] 由中国石油勘探开发研究院提供 (provided by RIPED)。

2022年 吉林油田累计注入 CO_2 达到 254 万吨[1]

2022: Jilin Oilfield injected a total of 2.54 million tons of CO_2 [1]

截至 2022 年 9 月 20 日，吉林油田开展驱油试验及示范应用超过 10 年，累计注气井组 88 个，注入 CO_2 总量达到 254 万吨，累计驱油 32 万吨。正在编制百万吨级 CCUS 示范工程实施方案，新建 20 万吨/年 CCUS 工业化示范应用项目也已经进入现场实施阶段（图 2.28）。

As of September 20, 2022, Jilin Oilfield had carried out CO_2 flooding tests and demonstration applications for more than 10 years, with a total of 88 gas injection well groups used, 2.54 million tons of CO_2 injected, and 320,000 tons of CO_2 flooded. It is now working out the implementation plan of a million-ton CCUS demonstration project, and its newly-built 200,000 t/y CCUS industrialization demonstration and application project has entered the field implementation stage (Fig. 2.28).

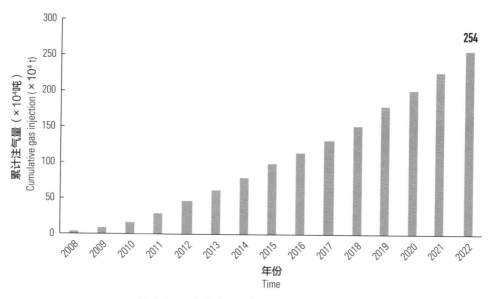

图 2.28 吉林油田历年来累计注入 CO_2 总量

Fig. 2.28 Total amount of CO_2 injected over the years in Jilin Oilfield

---

[1] 由吉林油田提供 (provided by Jilin Oilfield)。

2022年　辽河油田启动特低渗透油藏 CO_2 驱油埋存项目[1]

2022: Liaohe Oilfield kicked off a CO_2 flooding and storage project in ultra-low permeability reservoir [1]

2022年3月，辽河油田启动双229块洼128井区沙一段特低渗透油藏碳驱油碳埋存先导试验，计划3年内 CO_2 注入总量为508万吨。目前已启动产能建设和老井试注，并在杜家台古潜山等八个区块开展 CO_2 扩大试注试验，探索 CO_2 驱油技术在不同类型油藏适应性（图2.29）。

In March 2022, Liaohe Oilfield launched a pilot test of CO_2 flooding and storage in Sha 1 member in Wa 128 well area of Shuang 229 block, planning to inject a total of 5.08 million tons of CO_2 within three years. At present, production capacity construction and injection tests in maturing wells have been started, and expansion of CO_2 injection tests have been carried out in eight blocks including Dujiatai metamorphic buried hill to explore the adaptability of CO_2 flooding technology in different types of reservoirs (Fig. 2.29).

图2.29　辽河油田 CCUS 项目装置

Fig. 2.29　CCUS plant in Liaohe Oilfield

---

[1] 由辽河油田提供 (provided by Liaohe Oilfield)。

2022年 长庆油田开启鄂尔多斯盆地千万吨级 CCUS/CCS 产业基地建设新征程[①]

2022: Changqing Oilfield started the construction of a 10-million-ton CCUS/CCS industrial base in the Ordos Basin [①]

2022年9月,"长庆千万吨级 CCUS 中长期发展规划"和鄂尔多斯盆地首个百万吨 CCUS 工业化示范应用方案"长庆宁夏油区 CCUS-EOR 开发方案"通过股份公司审查,长庆油田正式开启鄂尔多斯盆地千万吨级 CCUS/CCS 产业基地建设。"十四五"期间,规划在陕、甘、宁三省区建成3个百万吨级 CCUS 工业化应用示范工程,形成617注1788采、每年 CO_2 注入320万吨的注入规模,支撑"十五五"末千万吨注入规模以及世界级 CCUS 产业集群的远景规划(图2.30)。

In September 2022, the "Medium-and Long-term Development Plan of Changqing Oilfield for Building a 10-million-ton CCUS Industrial Base" and "CCUS-EOR Development Plan in Changqing Ningxia Oil Area", the first million-ton CCUS industrial application and demonstration plan in the Ordos Basin, passed the review by CNPC. It marked the official start of the construction of a 10-million-ton CCUS/CCS industrial application and demonstration projects in the Ordos Basin by Changqing Oilfield. During the "14th Five-Year Plan" period, it was planned to build three million-ton CCUS demonstration zones in Shaanxi, Gansu and Ningxia provinces to realize annual CO_2 injection of 3.2 million tons with 617 injection and 1,788 production wells, serving as a support for the long-term planning of a 10-million-ton injection scale and a world-class CCUS industrial cluster at the end of the 15th Five-year Plan (Fig. 2.30).

图 2.30　长庆油田国家级 CCUS 先导试验基地

Fig. 2.30　National CCUS Pilot Test Base of Changqing Oilfield

---

[①] 由长庆油田提供 (provided by Changqing Oilfield)。

3 中国石油 CCUS 展望

3 Outlook on CCUS in CNPC

中国石油作为全球知名的能源公司和中国国有大型能源企业，深刻地认识到能源行业碳达峰、碳中和的重大战略意义。以 CCUS 技术为代表的负碳技术对实现"碳中和"目标具有决定作用，中国石油将积极推动 CCUS 技术尽早实现规模化、产业化，以便向用户和社会提供可信赖的负碳技术服务，实现 CCUS 技术从产业链向价值链的转变。为此，我们将：

As a world-renowned energy company and a Chinese state-owned energy enterprise, CNPC deeply understands the great strategic significance of carbon peaking and carbon neutrality in the energy industry. Since the negative carbon emission technology represented by CCUS technology plays a decisive role in achieving the goal of "carbon neutrality", CNPC will actively promote the scale-up and industrialization of CCUS technology as soon as possible, so as to provide users and society with reliable negative carbon emission technology services and realize the transformation of CCUS technology from the industrial chain to the value chain. To this end, we will:

加大 CCUS 超前技术研发力度，形成全球领先的 CCUS 技术体系。超前部署新一代吸附剂、高性能填料、中高温 CO_2 分离膜、新型捕集工艺、空气捕集等捕集技术，突破远距离大容量 CO_2 管输、封存、模拟、空天地一体实时监测技术，提升 CCUS 的经济性与安全性，参与全球直接空气碳捕获与封存（DACCS）、生物质能结合碳捕获与封存（BECCS）、海洋碳汇等研究合作，助力中国占领全球 CCUS 技术高地，实现本领域的高水平科技自立自强。

加快构建 CCUS 技术标准体系，促进我国 CCUS 的持续健康发展。积极开展 CO_2 捕集、运输、注入、封存等各环节技术和装备、安全环保风险监测与评估技术标准规范研究，构建 CCUS 全流程标准体系，推动我国完善财税激励政策和法律法规体系，探索建立 CCUS 定价机制与商务模式，深度参与国际标准体系构建，实现 CCUS 持续发展。

全力发展 CCUS 减碳负碳产业，打造碳循环经济示范点。发挥油田、炼化一体化业务优势，加强松辽、鄂尔多斯、新疆和环渤海等区域自有油田和炼化企业间合作，整合内部源汇匹配，构建四大盆地 CCUS 产业集群，形成完整减碳负碳产业链，推动建立适合我国国情的 CCUS 商业模式，助力我国碳达峰碳中和目标的实现。

加快 OGCI CCUS 区域产业中心建设，打造中国应对气候变化国家名片。继续强化油气行业气候倡议组织（OGCI）框架下全球 CCUS 区域产业中心布局与合作，在四大盆地等地区优先开展 CCUS 区域产业中心战略规划和建设，深化知识共享、技术输出和产业合作，深度参与国际多边机制有关活动，提升中国 CCUS 工作国际影响力和认可度。

Increase the research and development of advanced CCUS technology to form a world-leading CCUS technology system. CNPC will deploy a new generation of adsorbents, high-performance fillers, medium-high temperature CO_2 separation membranes and other capture technologies in advance, make breakthroughs in long-distance and large-capacity CO_2 pipeline transportation, storage, simulation, and air-space-ground real-time monitoring technologies, realize the capture, utilization and storage of low-concentration CO_2, participate in global research cooperation on DACCS, BECCS and marine carbon sinks, and help China occupy the global highland of CCUS technology.

Accelerate the construction of a CCUS technical standard system to promote the sustainable and healthy development of CCUS in China. CNPC will actively carry out research on technology and equipment in various links such as CO_2 capture, transportation, injection and storage, as well as technical standards and specifications for safety and environmental protection risk monitoring and evaluation, build a CCUS full-process standard system, promote China's improvement of fiscal and tax incentive policies and laws and regulations, explore the establishment of CCUS pricing mechanism, and realize the sustainable development of CCUS.

Vigorously develop CCUS-EOR carbon reduction and negative carbon emission industry and build a demonstration site of the circular carbon economy. CNPC will give full play to our advantages in the integrated business of oil field and refining and chemical industry, strengthen the cooperation between our own oil fields and refining and chemical enterprises in Songliao, Ordos, Xinjiang and Bohai Rim, integrate the internal source-sink matching, construct CCUS industrial clusters in four major basins, form a complete carbon reduction and negative carbon emission industry chain, promote the establishment of CCUS business model suitable for our national conditions, and help the realization of the goal of carbon neutrality and carbon peaking in China.

Accelerate the construction of the OGCI CCUS regional industrial center and create a national name card for China's response to climate change. CNPC will continue to strengthen the layout of global CCUS regional industrial centers and cooperation under the framework of the OGCI, give priority to the strategic planning and construction of CCUS regional industrial centers in the four major basins and other regions, deepen knowledge sharing, technology export and industrial cooperation, deeply participate in relevant activities of international multilateral mechanisms, and enhance the international influence and recognition of China's CCUS work.

附表 1　中国石油 CCUS 大事记一览表

序号	年份	大事记名称
1	1965 年	大庆油田碳酸水注入试验拉开中国探索 CO_2 驱油序幕
2	1984 年	大庆油田与法国公司合作开展 CO_2 驱油可行性及矿场试验研究
3	1994 年	吉林油田开展油田多井组 CO_2 吞吐试验
4	1999 年	吉林油田开展 CO_2 驱油试注实验
5	2003 年	大庆油田开展特低渗透储层 CO_2 驱油先导性试验
6	2004 年	中国寰球工程有限公司建成具有自主知识产权的 30 万吨 / 年中浓度 CO_2 低温甲醇洗分离技术
7	2006 年	中国石油在香山会议首次提出中国发展 CCUS/CCS 产业倡议
8	2007 年	中国石油承担国家 973 计划项目"温室气体提高石油采收率的资源化利用及地下埋存"
9	2008 年	中国石油承担国家科技重大专项"大型油气田及煤层气开发"下设项目和示范工程
10	2009 年	中国石油承担国家 863 计划项目"CO_2 驱油提高石油采收率与封存关键技术研究"
11	2009 年	中国石油重大科技专项"吉林油田 CO_2 驱油与埋存关键技术研究"启动
12	2011 年	设立"中国石油低碳关键技术研究"重大科技专项进行 CO_2 捕集与封存技术攻关和示范
13	2013 年	宁夏石化公司建成 15 万吨 / 年低浓度烟气 CO_2 捕集装置
14	2013 年	大庆油田建成 13.5 千米 CO_2 气态输送管道
15	2014 年	吉林油田建成十万吨级 CCUS-EOR 全流程示范工程
16	2014 年	大庆油田开展特低渗透油层 CO_2 非混相驱油工业化示范应用试验
17	2017 年	中国石油参与 CCUS 国际标准（ISO/TR 27915）制定
18	2017 年	南方石油勘探开发有限责任公司启动 CO_2 提高油气采收率与埋存先导试验
19	2018 年	中国石油参加巴西 CCUS 项目
20	2019 年	新疆油田启动 CO_2 混相驱先导试验
21	2019 年	首次在中国举办 OGCI 系列工作会议
22	2019 年	新疆 HUB 成为 OGCI 全球首批 5 个 CCUS 产业促进中心之一
23	2021 年	成立中国石油二氧化碳捕集、利用与封存重点实验室
24	2021 年	中国石油重大科技专项"二氧化碳规模化捕集、驱油与埋存全产业链关键技术研究及示范"启动
25	2021 年	成立中国石油碳中和技术研发中心
26	2021 年	中国石油担任全国 CCUS 标准工作组组长单位

续表

序号	年份	大事记名称
27	2021年	牵头编写的《OGCI中国CCUS商业化白皮书》正式发布
28	2021年	部署"四大工程示范"和"六个先导试验"CCUS工程,年注入能力超50万吨
29	2021年	大港油田实施深层低渗油藏CO_2驱油埋存井组试验
30	2022年	中国石油召开CCUS工作推进会,启动建设松辽盆地300万吨CCUS示范工程
31	2022年	新疆油田制订CCUS/CCS双千万吨产业发展远景规划
32	2022年	中国石油成立CCUS工作专班
33	2022年	提高油气采收率全国重点实验室获批建设
34	2022年	《中国石油绿色低碳发展行动计划3.0》发布
35	2022年	中国石油CO_2年注入量实现111万吨,累计注入量563万吨
36	2022年	吉林油田累计注入CO_2达到250万吨
37	2022年	辽河油田启动特低渗透油藏CO_2驱油埋存项目
38	2022年	长庆油田开启鄂尔多斯盆地千万吨级CCUS/CCS产业基地建设新征程

Appendix 1   Table of Chronicle of Major Events on CCUS in CNPC

No.	Year	Name of Event
1	1965	Carbonated water injection test in Daqing Oilfield ushered in a new era of CO_2 flooding exploration in China
2	1984	Daqing Oilfield partnered with French companies to carry out a CO_2 flooding feasibility study and field pilot test
3	1994	Jilin Oilfield carried out a multi-well group CO_2 huff-n-puff test in the oilfield
4	1999	Jilin Oilfield carried out a CO_2 flooding trial injection experiment
5	2003	Daqing Oilfield carried out a CO_2 flooding pilot test in ultra-low permeability reservoirs
6	2004	China Huanqiu Contracting & Engineering Co., Ltd. (HQC) built a 300,000 t/y medium-concentration CO_2 low-temperature methanol washing separation technology with independent intellectual property rights
7	2006	CNPC raised the initiative of developing the CCUS/CCS industry in China for the first time at the Xiangshan Science Conference
8	2007	CNPC undertook the Utilization of Greenhouse Gas as Resource in EOR and Storage It Underground, a project under the National 973 Program
9	2008	CNPC undertook a project under the Development of Large Oil and Gas Fields and Coalbed Methane, a national science and technology major project, and demonstration
10	2009	CNPC undertook the Study on Key Technologies of CO_2 Flooding for EOR and Storage, a project under the National 863 Program
11	2009	CNPC launched a major project "Study on Key Technologies of CO_2 Flooding and Storage in Jilin Oilfield"
12	2011	CNPC set up a major science and technology project "Research on Key Low-Carbon Technologies of CNPC" to explore and demonstrate CO_2 capture and storage technologies
13	2013	Ningxia PetroChemical Company built a 150,000 t/y CO_2 capture plant from low-concentration flue gas
14	2013	Daqing Oilfield completed a 13.5km gaseous CO_2 transmission pipeline
15	2014	Jilin Oilfield completed a 100,000-ton full-process CCUS-EOR demonstration project
16	2014	Daqing Oilfield carried out an industrial application test of CO_2 non-miscible flooding in an ultra-low permeability reservoir
17	2017	CNPC participated in the formulation of the CCUS international standard (ISO/TR 27915)
18	2017	China Southern Petroleum Exploration & Development Corporation launched the pilot test of CO_2 enhanced oil and gas recovery and storage
19	2018	CNPC participated in a CCUS project in Brazil
20	2019	Xinjiang Oilfield started the pilot test of CO_2 miscible flooding
21	2019	A series of OGCI working meetings held firstly in China
22	2019	Xinjiang Hub was selected as one of OGCI's first five CCUS hubs in the world
23	2021	The Key Laboratory of Carbon Capture Utilization and Storage, CNPC established

(continued)

No.	Year	Name of Event
24	2021	CNPC launched a major science and technology project "Research and Demonstration of Key Technologies for the Whole Industry Chain of Scalable Carbon Dioxide Capture, Oil Flooding and Storage"
25	2021	R&D Center for Carbon Neutralization Technology, CNPC established
26	2021	CNPC served as the leader of the National CCUS Standardization Working Group
27	2021	*OGCI CCUS in China Commercialization White Paper*, prepared under the leadership of CNPC, was released officially
28	2021	The CCUS industrialization project including 4 demonstration projects and 6 pilot tests with an annual injection capacity of 500,000 tons was pushed forward
29	2021	The storage well group test of CO_2 flooding in the deep and low-permeability reservoirs was carried out in Dagang Oilfield
30	2022	CNPC held a CCUS work promotion meeting, at which a 3-million-ton CCUS demonstration project in Songliao Basin launched
31	2022	Xinjiang Oilfield developed a long-term plan for the double ten-million-ton CCUS/CCS industry development
32	2022	CNPC established a CCUS working group
33	2022	State Key Laboratory of Enhanced Oil Recovery approved for construction
34	2022	*CNPC Green and Low-Carbon Development Action Plan 3.0* released
35	2022	CNPC achieved an annual CO_2 injection of 1.11 million tons, with a cumulative injection volume of more than 5.63 million tons
36	2022	Jilin Oilfield injected a total of 2.5 million tons of CO_2
37	2022	Liaohe Oilfield kicked off a CO_2 flooding and storage project in an ultra-low permeability reservoir
38	2022	Changqing Oilfield started the construction of a 10-million-ton CCUS/CCS industrial base in the Ordos Basin

附表 2　中国石油开展的 CCUS 相关研究项目

项目或课题名称	时间周期	项目来源
温室气体提高石油采收率的资源化利用及地下埋存	2006—2010	973 计划项目
二氧化碳减排、储存和资源化利用的基础研究	2011—2015	973 计划项目
CO_2 驱油提高石油采收率与封存关键技术研究	2009—2011	863 计划项目
含 CO_2 天然气藏安全开发与 CO_2 利用技术 / 示范工程	2008—2010	国家科技重大专项
CO_2 驱油与埋存关键技术 / 示范工程	2011—2015	国家科技重大专项
CO_2 捕集、驱油与埋存关键技术及应用 / 示范工程	2016—2020	国家科技重大专项
电—热耦合催化二氧化碳和甲烷制高附加值产物	2018—2023	国家科技重大专项
CO_2 加氢系列催化剂模式评价研究	2016—2021	国家科技重大专项
低能耗 CO_2 吸收 / 吸附技术工业示范和验证——15 万吨 / 年燃气烟气 CO_2 吸收示范装置工程方案设计	2017—2021	国家重点研发计划
高浓度 CO_2 捕集与地质封存技术集成和工程示范	2011—2015	国家科技支撑计划
深部咸水层二氧化碳地质储存关键技术研究	2012—2014	国土资源部公益性项目
含 CO_2 天然气开发和 CO_2 埋存及资源综合利用研究	2006—2008	中国石油重大专项
吉林油田 CO_2 驱油与埋存关键技术研究	2009—2011	中国石油重大专项
长庆低渗透油藏 CO_2 驱油及埋存关键技术与应用	2014—2020	中国石油重大专项
温室气体捕集与利用关键技术研究	2011—2015	中国石油重大专项
绿色油气田污染防治及生态保护研究	2021—2023	中国石油重大专项
大型氮肥工业化成套技术开发	2009—2019	中国石油重大专项
CO_2 法生产 DMF（N，N- 二甲基甲酰胺）工艺包	2020—2021	中国石油超前储备专项
天然气无氧催化转化及二氧化碳加氢高选择性制芳烃技术研究	2019—2021	中国石油超前储备专项
吉林油田 CO_2 驱先导试验 / 扩大试验	2009—2025	中国石油重大现场试验项目
大庆芳 48 区块和树 101 区块 CO_2 驱先导 / 扩大试验	2008—2025	中国石油重大现场试验项目
大庆海塔贝 14 区块 CO_2 驱先导 / 扩大试验	2009—2025	中国石油重大现场试验项目
长庆黄 3 区块 CO_2 驱先导试验	2017—2025	中国石油重大现场试验项目
新疆八区 530 区块 CO_2 驱先导试验	2017—2025	中国石油重大现场试验项目

Appendix 2 CCUS-related Research Projects Undertaken by CNPC

Project or Subject Name	Time Period	Project Source
Resource Utilization and Underground Storage of Greenhouse Gases for Enhanced Oil Recovery	2006—2010	973 Program
Basic Research on CO_2 Emission Reduction, Storage and Resource Utilization	2011—2015	
Research on Key Technologies of CO_2 Flooding for Enhanced Oil Recovery and Storage	2009—2011	863 Program
Safety Development of CO_2-bearing Natural Gas Reservoirs and CO_2 Utilization Technology/Demonstration Project	2008—2010	National Science and Technology Major Project
Key Technologies/Demonstration Projects of CO_2 Flooding and Storage	2011—2015	
Key Technologies and Applications/Demonstration Projects of CO_2 Capture, Flooding and Storage	2016—2020	
Production of High Value-added Products Based on Electric-thermal Coupling Catalysis of CO_2 and Methane	2018—2023	
Study on Model Test Evaluation of CO_2 Hydrogenation Series Catalysts	2016—2021	
Industrial Demonstration and Verification of CO_2 Absorption/Adsorption Technology with Low Energy Consumption—Engineering Scheme Design of 150,000t/y Flue Gas CO_2 Absorption Demonstration Plant	2017—2021	National Key Research and Development Program of China
High-concentration CO_2 Capture and Geological Storage Technology Integration and Engineering Demonstration	2011—2015	National Basic Research Program of China
Research on Key Technologies for Geological Storage of CO_2 in Deep Saline Aquifers	2012—2014	Special Fund for Public Welfare Industry of Ministry of Land and Resources of the People's Republic of China
Development of Natural Gas Containing CO_2, Storage of CO_2 and Comprehensive Utilization of Resources	2006—2008	CNPC Major Science and Technology Projects
Research on Key Technologies of CO_2 Flooding and Storage in Jilin Oilfield	2009—2011	
Key Technologies and Applications of CO_2 Flooding and Storage in Low-permeability Reservoirs in Changqing Oilfield	2014—2020	
Research on Key Technologies of Greenhouse Gas Capture and Utilization	2011—2015	
Research on Pollution Prevention and Ecological Protection of Green Oil and Gas Fields	2021—2023	
Development of a Complete Set of Technologies for Large-scale Nitrogen Fertilizer Industrialization	2009—2019	
Process Package for Producing DMF (N, N-dimethylformamide) by CO_2 Method	2020—2021	CNPC Advanced Reserve Projects
Research on Oxygen-free Catalytic Conversion of Natural Gas and High-selectivity Aromatic Hydrocarbon Production by CO_2 Hydrogenation	2019—2021	
Pilot/Expanded Test of CO_2 Flooding in Jilin Oilfield	2009—2025	CNPC Major Field Test Projects
Pilot/Expanded Test of CO_2 Flooding in Fang 48 Block and Shu 101 Block of Daqing	2008—2025	
Pilot/Expanded Test of CO_2 Flooding in Haitabei 14 Block, Daqing	2009—2025	
Pilot Test of CO_2 Flooding in Huang 3 Block of Changqing	2017—2025	
Pilot Test of CO_2 Flooding in Block 530, Area 8, Xinjiang	2017—2025	

附表3 中国石油CCUS主要示范工程

项目名称	捕集工业类型	捕集规模（万吨/年）	输送状态/方式	输送距离（千米）	用途	投运年份	2022年状态
吉林油田 CO_2-EOR 研究与示范	天然气处理	60	管道	20	EOR	2008	运行中
大庆油田 CO_2-EOR 示范项目	天然气处理	—	罐车+管道	—	EOR	2003	运行中
新疆油田八区530井区东部克下组砾岩油藏 CO_2 混相驱先导试验	甲醇厂	10	液态/罐车	33	EOR	2017	运行中
长庆油田姬塬黄3区 CO_2-EOR 项目	煤化工	20	液态/罐车	130	EOR	2017	运行中
大港油田叶21断块二氧化碳混相驱井组试验	—	—	—	—	EOR	2021	运行中
宁夏石化公司15万吨/年烟气 CO_2 捕集装置	烟气	15	—	—	生产尿素	2013	停运

Appendix 3 Major CCUS Projects of CNPC

Project Name	Capture Industry Type	Capture Scale (10,000 t/y)	Transport Status/Mode	Transport Distance (km)	Purpose	Year of Operation	2022 Status
CO_2–EOR Research and Demonstration in Jilin Oilfield	Natural gas processing	60	Pipeline	20	EOR	2008	In operation
CO_2–EOR Demonstration Project in Daqing Oilfield	Natural gas processing	—	Tanker + pipeline	—	EOR	2003	In operation
Pilot Test of CO_2 Miscible Flooding in the Conglomerate Reservoirs of Kexia Formation in the East of Block 530, Area 8, Xinjiang Oilfield	Methanol plant	10	Liquid/tanker	33	EOR	2017	In operation
CO_2–EOR Project in Huang 3 Block of Changqing Oilfield Jiyuan	Coal chemical industry	20	Liquid/tanker	130	EOR	2017	In operation
CO_2 Miscible Flooding Well Group Test in Ye 21 Fault Block of Dagang Oilfield	—	—	—	—	EOR	2021	In operation
Low CO_2 concentration flue gas capture equipment with 150,000 tons/year processing scale in Ningxia PetroChemical Company	Flue gas	15	—	—	Production of urea	2013	Suspended

在保护中开发
在开发中保护
环保优先